On the Network

On the Network

Jason Rakers

iUniverse, Inc.

New York Lincoln Shanghai

On the Network

iUniverse books may be ordered through booksellers or by contacting:

iUniverse
2021 Pine Lake Road, Suite 100
Lincoln, NE 68512
www.iuniverse.com
1-800-Authors (1-800-288-4677)

ISBN-13: 978-0-595-36494-7 (pbk)
ISBN-13: 978-0-595-80927-1 (ebk)
ISBN-10: 0-595-36494-2 (pbk)
ISBN-10: 0-595-80927-8 (ebk)

Printed in the United States of America

For Colleen and Charlie, may their love never cease to amaze me.

Contents

Introduction 1

CHAPTER 1 A Narrative History 5
- *The Past* *5*
- *The Present* *7*

CHAPTER 2 Ride the wave 10

CHAPTER 3 Exploring Information Technology
Environments 16
- *Academia* *17*
- *Banking* *18*
- *Electric Utility* *19*
- *Internet Service Provider* *20*
- *Independence* *22*

CHAPTER 4 Secrets of a Network Engineer 24
- *Customers* *24*
- *Learning* *25*
- *Technology* *26*
- *Knowledge* *31*

CHAPTER 5 Only time shall tell 36
- *The Future* *37*

References 41

"The development of an industrial society depended on capital as the critical resource. The most competitive automobile plants were those backed by the capital necessary to stay at the edge of industrial technology, and not many people had such storehouses of funds."

—Great Ideas in Management, by W. Jack Duncan

Introduction

Welcome to the real world. This is not the sugarcoated image of the world we see at a glance, but the bits and pieces that actually make our world work. Technology is reshaping our world into a new "high tech" universe where we no longer count in days or weeks, but in fractions of a second.

In 1980, the book *The Third Wave* by Alvin Toffler, provided an examination of three waves of society and the hidden connections among politics, business, education, and social life that influence a new generation of thinkers, believers, educators, and leaders.

I first encountered *The Third Wave* as an undergraduate student in college. My honest first impression of the book, "Wow, this book is long!" My second impression, "Wow, this book is amazing…when was this thing written?" That's when I realized over a decade had gone by since this book was published, yet I was now living in a society that many times resembled this third wave. If I look to a point in my life that changed my perspective, this was it. From that day forward I have pursued a life in what has now been called Information Technology. Information Technology can be described as the support and use of technology solutions to gather and share information. I am more than simply riding the third wave I am living it.

The quote from W. Jack Duncan's book *Great Ideas in Management* refers to the fuel behind the success of an industrial society. This is an important concept to consider, for without the deep pockets of capital resource financing the industrial revolution may have stalled. For the Internet generation the information age is fueled by innovation, free will, and the human resource. The Internet is like oxygen to a flame, it fuels the open and free exchange of information among people. Stifle the oxygen and the flame of the information revolution grows dim.

For the purpose of this book, Information Technology (IT), Information Systems (IS), and computers, or computer technology, are used interchangeably. I make no attempt to distinguish one from the other. Some would argue these words are not synonymous and are very different beasts, to which I agree. However, for the intended audience of this book these terms can be considered the same. I classify them as equally the same for the basis of perspective. Many people consider computers to be technology and technology often implies the use of a computer, or some sort of CPU. While I believe this perspective may be misleading and not entirely true it is not necessarily false either. As for IT and IS, they are primarily industry terms which now reach far beyond their original intended audience.

If the third wave is the information age, then it is an age where information is power. A power that is able to make or break the success of businesses, universities, and governments. The daily life of the information age includes an undeniable presence of pagers, wireless phones, mobile computers, and personal digital assistants. A dominating factor in helping the information age gain a lasting foothold in our lives is the growth of the Internet. The Internet has moved beyond a tool for universities to a way of life for millions and plays an ever increasingly important role in spreading information to the masses. Fifteen years ago the thought of sending an electronic message to friends and family rather than a traditional mailed postal letter was unthinkable, today the reverse would be true. We are merely at the beginning of a new way of life.

Whatever the name, Computer Systems, Management Information Systems, Expert Systems, or Information Technology they are the technology that empowers us to better our lives by working hand-in-hand with other systems to share information. Sharing, the thing children are taught to do at an early age and soon forget in their adult lives. Sharing information brings the world together. Information is important to society and sharing information is what makes the impossible seem merely improbable. Yet, information is only useful in the timeliness it is received. The ability to make information available in real-time is where the communications network plays a key role in making information, not only valuable, but also valid. It is the communications network that lets Hong Kong communicate with Bismarck, North Dakota in mere fractions of a second. The network is the backbone of the information age.

The network is the real story. It spans hundreds and thousands of miles: beneath the deepest blue sea, high in the stars above, and across the most treacherous terrain. A network is the infrastructure and backbone to the entire information wave. It is the bringing of information right to the point of contact where the network makes itself known. A society dependent on information is tied to the strength of the network on which the information travels from source to destination and back again.

Just as industrial nations rely on a well-maintained interconnected transit infrastructure of roads, railways, and highways, a society driven by information requires a well-maintained interconnected communications network infrastructure of copper, fiber optics, and wireless services. Recent years have seen telephone companies, telecommunication companies, and even utility companies building an infrastructure of fiber optic cables coast to coast to support the increase in demand. It is the successful use of the network that makes information valuable. It is the network that makes the Internet work. It is the network that brought the information age a little closer to home.

Welcome to the real world. A world very few ever experience. A world very few ever acknowledge. This is the world in which I live, the world of the Network Engineer.

This book, in its simplest form, is my attempt to explain my profession, my world, and my life. On countless occasions I have been asked from technologically challenged friends and family, "What do you do?" Each time as I try to explain, "what I do", I see the look of confusion when I say, "I work with networks. Networks allow people in a building talk to people in another building. Networks allow people, separated geographically, to communicate and share information." In desperation I simply say, "I work with computers!" My hope is the reader will gain an introductory understanding of what Network Engineering is about as a profession and the role of the network in the information age.

1

A Narrative History

Congratulations! You have taken the first step towards learning about the world of networking and Information Technology. To get an idea of how far networking has come since the early days, I present a fictitious example for us to explore. It is important to watch how information has been handled, what information was considered, and how new technologies changed the importance of information while forever impacting our world.

THE PAST

Joe was the manager of the Data Processing group for Corporation America in Chicago, IL. The key to data processing was stored on the appropriately named key cards or punch cards. These cards are paper cards keyed a certain way to be recognized by the computer as certain instruction sets. The actual data was contained on large spools of magnetic tape. All of the data was housed in a large glass enclosed room, nicknamed the glass house. In the glass house the ambient temperature was as cold as an early winter's morning. Within these glass walls stood the supercomputers of the day.

The main computer, the mainframe, processed all the data that mattered to Corporation America. The mainframe computer was the foundation for data processing. Without the mainframe, the key cards and magnetic tapes were useless. The mainframe processed millions of bits of data a day and Joe continuously changed key cards and spools of tape to keep pace with the endless demands of the machine. As times changed, so did the amount of data being processed and how it was processed.

In the distance, just barely over the thunderous noise of the data storage devices, Joe heard the faint sound of a telephone ringing. On the other end of the telephone was Sally. Sally works in New York for Corporation USA. Corporation USA and Corporation America had recently formed a strategic business alliance. The two companies decided they could better identify their current and potential customers by sharing data.

Joe had worked with Sally prior to this particular telephone conversation. Every month she and Joe would mail each other new sets of magnetic tape to be processed by their respected mainframes. Sally was just calling to request another spool of tape be sent to her. Joe assured her it would be in New York by the end of the week. Sally was quite surprised, for it seemed very quick, she was used to a two-week turn around time. Joe reminded her of the new courier service that went into effect last week between their two companies. After gathering all the mail for each office of Corporation America, the courier drove it directly to the appropriate corresponding Corporation USA office, and then repeated the process for Corporation USA.

In the beginning, data was data. Data was stuff and mattered after a decision was made. Data was used for verification of decisions, inventories, and other key elements. Today, data is still data. Data is still stuff, but now data is interpreted into useful information and typically used prior to making decisions. Decisions and inventories can be based on real-time information, rather than historical information. Modern technological advancement is the basis for the JIT (Just-In-Time) inventory. Using a JIT inventory allows organizations to reduce overheads by replacing components on-demand, rather than by an inventory of stored components.

Information has become more than important to the survival of companies, organizations, people, and governments; it has become critical. The traditional courier service transporting data over the sneaker-net is no longer fast enough, reliable enough, or practical enough. The sneaker-net is a description of a network based on the movement of physical data between computers. For instance, saving a file from one computer onto a floppy disk and then walking in one's sneakers to another computer to use that file. A courier service moving data between locations is another example of a sneaker-net in action. This sets the stage for the present. Let's take another look at our friend Joe.

THE PRESENT

Joe has been working for Corporation America for over twenty years. He has seen the progress of technology all around him. Joe's glass house has become the home for numerous other computers mainly performing specialized tasks, such as Application Servers, Web Servers, Database Servers, and Domain Name Servers. Yet, none of these servers have been able to effectively replace Joe's trusted mainframe. As technology has changed over the years so has the mainframe, but it is still typically the biggest and most reliable system in the house.

Today, the data is no longer stored on spools of magnetic tape with instruction commands issued from key cards. Now data is stored on CDs, DVDs, miniature tapes, storage area networks, and removable hard drives all receiving commands from multi-tasking operating systems. The glass house is still glass, but the ambient temperature now resembles a cool spring day. Joe still occasionally changes tapes or disks, but mainly the mainframe runs without the need for someone to standby and feed it data. In fact, the majority of the data would appear to arrive magically! The mainframe has a cable attached to it that allows the mainframe to be connected to other computers over what is called a network. This Local Area Network (LAN) cable passes billions of bits of data to and from the mainframe in a matter of seconds.

In the distance, over the rumbling noise of all the other devices in the glass house, Joe hears the faint sound of a cordless telephone ringing. On the other end of the telephone is Sally from Corporation USA. Sally calls Joe daily to confirm the data has arrived from New York properly, without any courier, but through the network cable now attached to the mainframe using what is called the File Transfer Protocol (FTP).

Sally complained about the FTP taking nearly 60 minutes today. Joe reminded Sally that the file transfer always takes a little longer on Fridays, because of the additional data needed for end of week processing.

FTP is a means of transferring data files from one computer to another using a protocol, or language, called the Internet Protocol (IP). A protocol is what helps computer systems communicate with each other. There have been many different protocols over the years like the Internetwork Packet Xchange, Systems Network Architecture, and AppleTalk. Typically protocols have been propriety, meaning a particular company created the protocol and only that company's computers (or products) could speak the protocol. The exchange of information between an Apple computer and an IBM mainframe was difficult, if not impossible. IP has become a technology industry standard across computer systems. IP allows computers on the same connected network, or other interconnected networks, to talk together efficiently and effectively.

Networking and Information Technology has come a long way. Thanks to the Internet and other advancements in technology, companies from all across the planet can communicate as if they are right down the street. Networking has made the acceleration of technology possible through the sharing of ideas and information. In return, technology has made networking faster, more reliable, and practical as well as made the career of the Network Engineer challenging and rewarding.

Networking touches many aspects of Information Technology. It allows the things we take for granted today to be possible. The ability to communicate with almost any device on the Internet is possible because thousands of networks are all interconnected. Networking even plays a role in defining the relationships humans have with their computers as friends, enemies, or even slaves.

Some would suggest humanity has become a slave to technology. However, I propose that it is humanity that has enslaved technology. It is not a man or woman making a fresh pot of morning coffee, but a machine. It is not John Deere cutting our lawns, but a self-sufficient solar robot. It is not a child learning to read, but an intelligent book reading to a child.

The new emergence of importance on security, privacy, and accountability are all results of the ability to internetwork once unrelated systems. Organizations, such as businesses and governments, have had to rework old models of interaction to exploit new opportunities and plan for potential threats from competitors.

We will examine some of the more intriguing aspects of the impact of networking on Information Technology including the apparent presence of computers everywhere we look and the relationships people share with technology. First, we prepare ourselves by riding the wave of change.

2

Ride the wave

Surf's up! Change has become an all to common occurrence in our modern times. We can often become overwhelmed and feel as if we are drowning in the change around us. *The Third Wave* details Toffler's description of the First, Second, and Third Waves with the waves roughly representing different periods of life: agriculture, industrial, and information. I think the use of waves is an appropriate metaphor to describe change. As the threatening waves of an ocean range in size from the timid to the untamed, so too can the essence of change vary in its cause and effect.

I use the analogy of waves hitting the shores of a small island to help us picture what is occurring in our world. When a wave of change comes, it is just like the initial waves from a gathering storm on the horizon coming in from the sea. Often the waves are small and merely tempt us at something bigger to follow. The waves grow and dissipate, fade and contrast, often without our knowing. But if we heed the warnings we realize a much larger event is swirling beneath these oceanic tides. Suddenly it appears, the overwhelming presence of a typhoon and cleansing nature of the tidal wave.

The people and surrounding regions near the shoreline become engulfed by the incoming wave. People often try to resist the tidal forces. The actions of sandbagging shorelines and other attempts at building barriers between oneself and the looming waves are all in an effort to postpone, or more hopefully prevent, the ravaging effects. Yet, in the end the unstoppable force of Mother Nature plays her hand often leaving many victims in her wake. Those lucky enough to escape the ocean's wrath still feel the impact left by its mere presence. There are those people farther inland from the shoreline that never experience the full force of the wave, but see the ripple of effects resulting from these chaotic events. Finally there are those who do not fall victim to the crashing waves merely because of how well entrenched in existing barriers they chose to exist.

Change is like this wave of water, for in change we are often overcome by the new reality presented to us. Change results in a new perspective on how things are done, accomplished, and achieved. Living the lifestyle of an agricultural based society is drastically different than the lifestyle of an industrial nation. Now compare the difference between both of these lifestyles with that of the information based society. Perspectives are modified and new paradigms are created.

Change is not limited to our social environment. Change is very much a part of technology and the evolution of technological advancement. Microprocessors get faster, smaller, and more cost effective with every new achievement. Networking topologies migrate from shared to switched to shared again. The life of an IT professional requires a constant refresh to the surrounding environment. In networking the waves of change have impacted how organizations are interconnected.

When the mainframe ruled the world, the data network was a strict hierarchy with the mainframe at the center and the client, or host terminals, at the edge. We know the mainframe was the brain to all operations so vital life lines were required to interconnect these logical units in the field back to the physical unit in the classic datacenter. However, client-server architectures began a systemic revolution of freedom for the dumb terminal masses. For the first time these dumb terminals were free from their thought-dominating masters.

The client-server architecture sought to push more intelligence to the desktop. The enlightening of the dumb terminal and empowering it to make decisions on its own opened a door of opportunity for those willing to explore. The resulting freedom of information processing required the network to change. It had to move beyond the classical strict centralized hierarchy to a distributed mesh of interconnected and intelligent devices.

The earliest of these new network forms used a shared-bus topology to interconnect devices. All network-attached nodes shared the same physical bandwidth available. This was a very effective design for small networks. However, as networks grew so did the amount of traffic needing to be transported across the network. The unfortunate side effect of success resulted in resource contention and poor performance for all connected devices. A new direction was required.

With the rapid adoption of client-server architecture for its ease of use and liberating move to graphical user interfaces, network contention became a serious problem for users on the shared topology. Advocates of the hierarchical mainframe generation called this a major draw back to client-server endeavors, and another reason why the mainframe way of existence was the right way. These stubborn resistance leaders refused to give into change. They had barricaded their doorways and buried their balding heads into the sand in a lasting attempt to resist. The two worlds of networking technology could not coexist. So much the same, yet the mainframe and client-server were so very different as well.

As any good gene would do, the shared network topology found an answer to its dilemma through mutation. The answer came with the adoption of a switched topology.

Switched networks allowed for the control of contention on the network by regulating collision domains and allowing the formation of multiple broadcast domains. These small modifications allowed the growth of radical client-serverism to continue its propagation beyond the historical shared topology restrictions. Switched topologies opened the door to additional growth and more network intensive application development.

Users of switched topologies looked upon their shared topology cousins as the unwanted bastards of their past. Shared topologies were a constant reminder of the flaws inherent in the initial client-server movement. The two would forge ahead and learn to live together in peace. The mainframenites stood in awe as the limitations of the client-server architecture fell to the wayside. Flat networks were surely to dominate the world forever shattering the mainframe hierarchy of control. Then again, change was knocking on the door.

There is an old adage about rocking and knocking that has merit. However, in the client-server architecture the return of an old nemesis meant something more positive. Switched flat networks were a hot property, but the introduction of a hierarchical switched environment made heads spin.

In this hierarchical network, the network is comprised of three distinct entities: access layer, distribution layer, and core layer. The model can be loosely applied to the centralized approach of the mainframe. The access layer represented the client access to the network, often comprising many connections. The distribution layer provided intelligent switching and routing of network traffic to various other access layer or distribution layer components. The core layer provided fast interconnection among the distribution layer elements. If the mainframe and dumb terminal represented the distribution and access layers respectively, then the core layer provided the interconnection among many unique mainframes and their associated dumb terminals.

The adoption of the hierarchical network design fueled the growth of network environments. The 80/20 rule of networking refers to 80% of traffic remains local while 20% of traffic is remote. The remote traffic would be routed among distribution layer elements while local traffic remained in the access layer. As the past had dictated, however, the continual change of technology innovation is constant.

The new network rule is the 20/80 rule. This rule states that 20% of traffic is local and 80% is remote. The conflicting rules are indicators of the change from the client-server distributed environment to a centralized client-server environment. The introduction of thin-clients and remote terminal service applications removed the brains from the access layer devices and restored them to centralized application servers. There and back again a full circle we make, well almost.

When a network engineer or any other IT professional realizes the implications of technology and its role in helping businesses achieve competitive advantages over competitors it should be obvious to consider not only the present, but also the future role technology will play. Successful implementations of networking, application, or server-based solutions will not remain successful as the environments of use change. We ride the wave of change. If we are good we stay the course, if we are not so blessed we drown in a sea of disparate and uncommunicative technologies.

If all of the waves of change have influenced our lives on such a tremendous scale, then how are we to identify success? Our belief in politics and economics or family and education has been shaped. Our acceptance of the Internet as a way of life, or the MP3 as the music format of choice has been shaped. Everything has been changed by the evolution of the past and everything will be changed by the revolutions of the future. We face a time of change. We must embrace the change for all its worth, or resist it by holding on to our old ideals. Today requires a new way of thinking. Information, technology, and services are opening our eyes to the rest of the world and the future of what could be possible. The future is in the eye of the beholder, the eyes of the network engineer and other IT professionals, not Microsoft, Cisco, or IBM.

The future belongs to the individual. It is the power of the individual, the power of one, to make a difference. Do I build the right solution or the wrong solution? Do I help my neighbor when they fall down? Do I close my eyes and look the other way to the harshness this world often brings? We all have the power to do what is right.

Our future is the future we make it. The power of the individual to try, to love, and to dream. To become who we should be, not just another drone. Success is not difficult if we have the right tools. To achieve success, we only need the strength, courage, and will power to succeed. We must ride the wave.

3

Exploring Information Technology Environments

Let us take a moment now to examine some of the different types of IT environments. My life as a Network Engineer has crossed the institutions of higher education and academia, banking, public utility, Internet Service Provider, and the ever-popular role of consultant. In each of these arenas I have had the pleasure of working with a multitude of talented, yet uniquely different, individuals. Just as my colleagues have differed in each environment so too has the working environment itself.

As you will read shortly, the environments range from the tame to the wild and absurd. There is no sure way to determine what environment is the best for a person to work. It is obviously dependent upon an individual's preferences, a strictly subjective matter of opinion. Environments can be radically different based on the organization itself, rather than the industry as a whole. Some people like a strict and rigid work place while others like a fly by the seat of your pants work place. The following are based on my research, observations, and personal experiences within each of these environments.

ACADEMIA

If you are looking for a place to truly define who you are, then look no farther than the institution of education. This institution of higher learning was a small private college. It offered the ability to explore and learn while helping to define a skill set and area of expertise. The atmosphere was typically relaxed. People for the most part were there because they truly enjoyed their professions. If you are looking for a highly rewarding job financially then education is not the place for you, but if you are looking for a different type of rewarding opportunity you found a great spot to land. Your financial compensation may be low compared to what a profit motivated organization would pay, but the opportunities and environment more than make up for almost any of the financial differences.

The world of higher learning is very open-minded to new ideas and paradigms on the way things should be done. There is a certain amount of freedom in the management of the organization. Most importantly, education institutions openly encourage employee growth and follow-up their words with actions. Depending on the extent and scope, you are often able to make decisions without the need for additional approval. Perhaps this is why security always seems to be lacking from the infrastructure and the daily lives of its people.

IT infrastructure is a second-class citizen, but only because technology is neither a means nor an end. Often the technology is treated as a necessary evil, which surprising contradicts the growth towards laptops in the classroom and online learning. Good equipment is usually hard to come by and equipment that is procured often requires a special touch of inventive flair. Thus, a just make it work philosophy is often required.

Education seems to be the place everyone wants to work. Often times however, they realize the green pastures of their dreams are not very green when they finally stand in the middle of the freshly mowed pasture. It can be a very difficult and barren place filled with its own special blend of cow pie surprise. Those who choose to remain are there because they want to be there, not because they need to be there.

Maybe the college campus is not the place for you. How about a place where all the money is located? Health care? No, you are not paying attention. The bank is where all the money is located. It is the economic depository for people all across the globe. It is the economic trading floor of business-people, homebuyers, college-savers, and retirement-takers.

BANKING

This financial institution, a small national bank, was a good place to work. In my experience the people were great to work with on a daily basis. They had an honest appreciation for ensuring the company's success. The company treated its technology personnel well and as respected members in their unique fields. The environment was serious with a special attention to details. Unlike universities, mistakes here usually cost millions of dollars in potential revenue. Technology planning has a high priority with thought-out decisions and a well-developed project management structure.

Financial institutions tend to be close-minded to new ideas and very slow to react to meeting customer demands. Management can be rigid in decision making which tends to result in slow decisions. Security, however, is of the utmost importance and is taken very seriously. Even jokes can often times lead to serious reprimands. Equipment is usually available and you are encouraged to build solutions not only cost efficiently, but solutions that work right the first time. Employees are often encouraged to grow, but only based on the success of their achievements.

The notion of recognition based on achievement is not necessarily a bad thing. It helps to keep resources on top of their game and ensures a continual focus on success. However, it can be very frustrating when your projects just do not seem to yield success.

Alas, the house of coin and paper bill is still not your cup of tea? Well, how about the workhorse of every major industrial nation? The electric utility powers the globe. Perhaps the most important requirement in the information age is the need for electricity. Without power, computers would make the best desk ornaments the world has ever seen.

ELECTRIC UTILITY

If you were looking for a place where your career success is measured only by your most recent success, you found it at this regional electric utility. Here people are treated as true resources. These resources may be human, but again resources nonetheless. Everyone was a resource easily replaceable with another human resource. The idea of being skilled in a specific area and focusing resource talents in this area never seemed to reach upper management.

This particular organization had implemented a poorly designed and tragically implemented team structure. Without any one manager to provide guidance the technology teams were left to their own to maintain the technology needs. The organization then followed up this approach with a haphazard centralized project management implementation. The saying too many chiefs and not enough Indians was experienced first-hand.

Many issues often went unresolved based on a strict adherence to the project management philosophy. In order for any work to be accomplished an accompanying project must receive sponsorships from management. Yet, not everything in technology support can be summed up into a nice and neat project with clearly defined start and end dates and yield packaged deliverables. Many information technology activities are daily, or ongoing events. This is the nature of the beast in Information Technology. Anyone familiar with technology should understand this component requirement. I have been a part of successful team environments before, but here it had failed.

There can be a very close-minded approach to new ideas and any forms of improving efficiency. Management can be rigid and intent on maintaining the status quo. There often is a strong belief in the motto, "if it isn't broke, don't make it better." IT Security was obscure, it was constantly mentioned and seen, but never used appropriately. Equipment was widely available and usually the best of the breed. There seemed to be no limit to the budget for technology resulting in haphazard implementations all too frequently. Employees were often lost in the daily grind, typically reporting to numerous supervisors. It was sad, but true, that employees were often only recognized for their achievements when the monthly e-mail arrived notifying current employees of the recent employee and retiree deaths. Luckily this organization happened to realize its mistakes.

It evolved through several reorganization attempts into a champion for first-rate organizations. This organization made significant steps in improvement and deserves the recognition. Yet, change is inevitable. If an organization does not succeed it fails. If I take this approach with all organizations reviewed I would never complete this book. I think it would be an accurate assessment to state if this organization did not change when it did, it would no longer exist.

If we learn anything from this chapter, it should be how organizations do change over time. Sometimes they change abruptly, but more so the change is subtle and we do not realize it until we reflect on the past.

INTERNET SERVICE PROVIDER

The pinnacle of the dotcom generation was the explosion of the Internet Service Provider (ISP) market. Without a doubt this environment sparked the most interest for techies and non-techies alike. For techies the ISP was the place to escape from the traditional ho-hum existence of corporate America's technology support. All geeks could feel at home within an ISP. They were given the freedom and challenges desired in their daily work lives. As for the non-techies, working for an ISP meant increased financial rewards. Unfortunately, this proved to be the downfall for many dotcom era organizations. ISPs offered huge compensation packages including unheard of bonuses for technical and non-technical staff alike.

This Internet Service Provider was a regional ISP with a very active imagination of opportunities for profit. For those requiring a translation, this meant management had solid delusions of grandeur regarding the future success of the company. It was believed they would gain 6,000 new customers within the first six months of opening the doors. Management believed the marketing hype spreading across the globe on the future of the dotcom generation. If you build it they will come did not quite reach fruition.

As was common with all Internet Service Providers, security was not a concern for Internet access. In fact, security was often seen as an obstruction to new markets. The Internet, after all, was about freedom: the freedom to communicate and the freedom to be unknown. In many ways the Internet was like the proverbial emperor's new cloak. As users of the Internet, we are naïve to believe that our beautiful new clothing will allow us to be something more than ourselves, as if we held a shield of invisibility. In reality the shroud of our ignorance left us visible and exposed to a world eager to dismantle our lives.

In summary, the Internet Service Provider was a child of the once new era of technological revolution. It offered an open-mind and active imagination for well-treated employees and management, and the promise of the Technicolor rainbow kept everyone focused on the pot of gold at the other side. The budgets of the ISP seemed limitless with the deep pockets of venture capitalists. Unfortunately, the pot of gold was only a cup. It seems someone incorrectly converted from the metric system.

INDEPENDENCE

The last institution is an American classic: the consultant. In the 1990s everyone wanted to be a consultant. Not just any consultant, but an independent consultant. Independent meant freedom from the confines of corporate America and the structure of corporate political intrigue. Yet independent consultants were often left to fend for themselves. The endless pressures of ensuring new business opportunities were available at the end of a current engagement left many exhausted. Consultants working for the traditional consulting company on the other hand, receive the benefits and financial wealth without the hassle. Consultants play by a different set of rules where cubicles do not exist and paychecks are fat.

To almost anyone the life of a consultant is exciting, at least from a distance. If you have been fortunate to work with a number of consultants you probably have noticed they do not quite know what they are doing exactly. By its very nature the role of the consultant is ad hoc. In other words, we make things up as we go. Blasphemy! No, not really, it is honesty.

Often consultants are thrown into an engagement with a customer without the qualifications requested by the customer. The consulting agency plays the odds that the customer will not realize the inadequacies of the consultant. The odds often favor the agency since technology consultants can learn as they go through the life of the project. If the typical customer does not have the experience required for the project thus the need for a consultant; chances are good the customer will not realize the consultant is as clueless as the customer. In other words, the customer will not realize the consultant is actually ignorant since the customer is even more ignorant about the subject matter.

However, every once in a while the customer has the fortunate glimpse at the truth. After repeated glimpses you soon realize the consultant is not about helping you succeed. His or her goal is making money. People will pay exorbitant amounts of money to be told what to do and how to do it. The monetary payoff is a very lucrative and rewarding opportunity.

In truth the consultant is a great opportunity for learning different technology areas. The environment does not get boring and can always be challenging. If you are a good consultant then you are literate in the subject matters in which you are consulting and add value for your customer. Good consultants can find it hard to fit into a consulting agency environment, and often choose to work independently. If you can tolerate the misdirection and illusions often considered the norm being presented to the customer then consulting agencies are rewarding choices.

If you consider information technology to be your job, you will never truly be successful at it. Information technology is a way of life beyond any organization. Organizations come and go, but the technology that drives society forward remains. To be the best, your life becomes deeply involved with information technology. Technology changes too quickly to think you can be an expert by working eight hours a day in your field. If you are still wondering how amazing the life in networking can be then keep on reading and I will share some of the secrets I have learned from my experience. For those of you who care less about the ramblings of wisdom from an old coot you may skip to the next chapter.

4

Secrets of a Network Engineer

This chapter of the book identifies some experiences from my life as a Network Engineer. It is meant as a diversion for us from the world beyond and has roots in the present. I have decided to share these experiences in the hopes that some of the common pitfalls in the life of a Network Engineer may be avoided, or at least realized and joked about.

The secrets can be compiled into four areas of interest: Customers, Learning, Technology, and Knowledge. Each of these areas describes some of the lessons I have learned in my life as a Network Engineer. As we travel on this journey of the bizarre, insane, idiotic, and the other mind-numbing antics of networking please take the time to reflect and give thanks the world still turns round and that bits are not bytes. Enjoy!

CUSTOMERS

The customer is always right. It never fails that working in technology you will encounter people who claim to know everything and anything about a particular subject matter. Whether they are a peer or a client, it doesn't matter; it is still a pain to work with these people. By far, it is worse when the Supreme Guru is your client. You are obligated to be nice and professional to this person, who thinks reading a trade magazine article on Ethernet converters, makes him or her an Ethernet expert.

The key is not to let them get under your skin. You know what you are doing and you are the professional. Have confidence in your decisions and actions, but do not slip down the slippery slope of supreme guru yourself and become your peers' worst nightmare.

You will also discover that every client's application is the most important, or at least to the client whose application is not working at the moment. This is where being a professional plays it part. You need to be polite and demonstrate confidence by your actions to ease the boiling anxiety of your client. Politeness also pays off because not only will people enjoy being with you, but also you never know who may be your next boss in the reorganization of tomorrow.

My final thought on customers is to never burn your bridges. You never know when you may need to go back to an old job, position, or client. As the IT industry has demonstrated for years the demands of today can change with the wind. When the winds change you do not want to be left with nowhere to go.

LEARNING

A good outside organization can better understand your network than a bad internal staff; this is the law of outsourcing. A bad internal staff can be many things: lack of talent or skill, laziness, or a simple lack of commitment. A bad staff can make an outsourcing organization look very appealing to management. The important point is that no one will ever know your network better than a good internal staff.

A good internal staff is always touching the network; they understand how it functions, where the problems are, and the network's attitude towards Monday mornings. No outside organization can understand and maintain your network the way a good internal staff does. Outsourcing is more inclusive than the idea of contracting an outside organization to perform the internal IT functions. Outsourcing includes the role of the consultant.

Organizations hire consultants because they believe the organization lacks the skills internally for a particular objective. It never fails, however, that the truth is where you see it. Organizations hire expensive outside consultants to install new applications or design an information technology strategy. Yet these consultants seem to never quite understand what they are actually doing. Unfortunately the consultant is not sold based on skill, but rather marketing and the promise of success. Outside consultants never seem to know how to install the application you hired them to install.

Take a break. The best way to reenergize from a challenging morning is to go out to lunch with your coworkers at the local Chinese buffet restaurant. Mmm. All you can eat buffet! What else is there? And do not forget all fortunes inherently end with "in bed" at the end. It is a little well-known fact that the money saved by fortune cookie manufactures from not adding these six characters is over $150 million per year. Quite Amazing! Another amazing little known fact is some of the best network designs are created at the local bar, on cocktail napkins, around 11 o'clock in the evening.

Remember tomorrow is another day. Many times problems can become overwhelming and seem too impossible to solve. Do not give up. Perhaps all the situation needs is a new perspective. I have found that my best problem solving comes when I am sleeping. I am easily able to replay the events during the day and the steps I have taken in attempts to correct the problem. I am able to simulate new approaches in my head prior to implementing them, most times with great success the next day. Proper rest is essential for optimal performance. In the words of my father, the soccer coach, just "go for it!"

TECHNOLOGY

Well technology is what this book is about. Technology is the center of the universe, but there are several things that need clarification.

First, the acronym TCP/IP is everywhere nowadays. However, TCP/IP is not all there is to the puzzle.

IP, the Internet Protocol, has multiple transport layers; TCP, Transmission Control Protocol, is only one of them. In the seven layers of the Open Systems Interconnection model (OSI), the transport layer (layer four) is dependent on the network layer (layer three). With IP as the network layer protocol, the transport layer encapsulates IP packets before being passed onto the next layers of the OSI model. The two most common transport layers are TCP and UDP, or the User Datagram Protocol. Some other transport layers are Generic Routing Encapsulation (GRE) and Layer Two Tunneling Protocol (L2TP). Most likely people learned all they thought they needed to know about TCP/IP from any of a hundred industry trade magazines.

Unfortunately many people learn to design networks from trade magazines as well. All it takes is someone from management to read an article in a trade magazine on a hot new technology or how to make the Wide Area Network (WAN) better, and now the company needs this technology or solution. This process leads to a redesign every six months as new technologies are discussed in the magazines. Use trade magazines to keep abreast of industry changes, not for strategic planning.

Pagers are good. They keep you in touch with the network and its problems. It is important to maintain a sense of the environment. There was a time when only doctors carried pagers. However, nowadays it seems everyone and his or her mom has a mobile phone or pager. I guess we either have all become important, or there really is no escaping our lives. Of course every good has a bad.

Pagers are bad since they keep us in touch with the network and its problems. In an unstable network environment, the pager may become an unpleasant reminder of how much work still needs to be done in order to make your network better. If you have a pager you will undoubtedly agree. In fact it seems this darn little device knows the absolute worst time to go off.

Document is the one verb that will live on throughout the history of networking. Documentation may seem like a pain when doing it, but it can save your neck when there is a problem. Taking the time to document as much information as possible while installing equipment will save you time from having to document missed items later. After a few installs you are able to identify what is important, and what is not, making your documentation much more valuable. Keep a copy of the current configurations of equipment. When the equipment malfunctions, you will wish you had this type of documentation. Maybe your device configurations are simple and easy to implement, but when it is three o'clock in the morning you do not want to be reconfiguring a device from scratch by memory.

Since we are documenting you might as well follow the standard. Remember network standards should be followed, not just written in a document and forgotten about. If you develop standards they are most likely specific to your environment, rather than a generic industry standard.

When in doubt, read the manual. It is amazing what you can sometimes learn when you actually take the time to read the manual. Of course, many times it is also amazing how much you really cannot learn by reading the manual. Again documentation is key to your success. Document for your purposes so the knowledge is useful at a later time. If you use passwords on your equipment, learn how to circumvent and break into the equipment to recover a lost password before you actually lose a password. This will save you time when you need it most, and remember backdoors are doors of entry too.

Often we keep our enemies closer than our friends, the same is true in organizations. The worst security threat is someone working for the organization, perhaps Joe in Accounting. More security threats are located inside of an organization than outside. It is important to learn the traffic patterns of the network and be able to identify potential threats when they occur. The internal environment may be the worst threat, but the growth of the Internet has created an almost uncontrollable security threat externally.

It is important to not overlook the outside world from your security plans. While the Internet is beyond any single organization's control, there are simple measures that can be taken to help reduce the risk of external threats. Take caution when interacting with the Internet.

Take caution another step and remember that static electricity is bad. Be sure to ground yourself before working on equipment. Never work on the inside of equipment while there is power to it. Always turn off the power to the device as well as unplugging the device from its power source. Use a static guard to protect equipment from static electricity. Overlooking simple safety steps is a poor reason for potential damage to your equipment. In addition, technology requires power to operate and the use of power generates heat. However, a hot networking device is an unhappy networking device. Remember devices like the cool crisp breeze of a fall day. Do not overlook the environmental conditions, which are just as important as proper configuration and security of equipment.

When piloting new technologies, be sure your guinea pigs are aware they are part of a pilot, want to be involved with the pilot, and will actually provide you feedback during the entire length of the pilot. It is very awkward when a Network Engineer arrives one day to replace demo equipment with production equipment and the office staff thinks you are there to either fix the problem or throw the junk out. Pilot new technology whenever possible, prior to implementation, to ensure successful project results. When designing a network the best thing to remember is to keep it simple. Be smart in satisfying the needs of your client, but keep it simple enough to support so you will not be scratching your head in six months trying to figure out why anyone would design such a crazy thing.

Sometimes designs are not what we hoped, but having a bad design is better than no design. By having a light at the end of a tunnel to head towards will often be better than the alternative; stumbling around in the dark. Your direction may be off, but at least you have put thought into the design and it serves a purpose. A final important point in design and implementation is that purchasing networking equipment is not like going to the grocery store.

Ship dates for quality networking equipment can often be long, be sure to give yourself enough time when working on projects to order the equipment. A good rule of thumb is four weeks from order placed to product received. This also goes for ordering data circuits from a service provider. No matter how much time you plan for it always seems to take longer, and of course the actual install date the service provider gives you does not mean the circuit is ready...you still have to call them and turn-up, or activate, the circuit. A good rule of thumb for circuits is six weeks from order placed to turn-up date.

Cable modems are not really modems. Modems convert digital signals to analog frequencies and back again. Cable service is already a digital service, thus cable modems merely act as a gateway from the computer to the cable media. In addition there is a unique difference in the broadband abilities of cable and DSL.

Cable service Internet access is a shared service end to end, from the user's PC to the Internet. It is shared with everyone on your block, and perhaps more. Cable providers establish locations, called hubs. Each hub provides the cable broadcast (one to many) service for a certain geographic area, which is shared across the same media

Digital Subscriber Line (xDSL) Internet access is a shared service only from the providers Central Office (CO) to the Internet, and a private service from the user's PC to the CO. The Central Office is comparable to a Cable provider's hub, except that phone service has a unicast (one to one) relationship with customers.

Batch is dead. These are the infamous words of a misinformed colleague. Living proof that if you live by the trade magazine you die by the trade magazine. Batch file processing is the traditional form of data processing where commands are run on a periodic basis, usually every night. Information is not updated for use until after the batch process has been completed. Batch file processing still has its place. Today things may be handled more frequently in real-time, but batch jobs will remain a part of our lives for a long time to come.

Practice safe computing. Caution can help keep your computer and network healthy and secure, just as it can help keep you physically healthy and secure. Some examples of safe computing: do not go to untrusted websites, do not open e-mail attachments from people you do not know, do not open strange e-mail attachments from people you do know, and do not put strange floppies or other portable media in your computer. It is best practice to use anti-virus software and to close any opened or unnecessary ports/applications on your system.

Safety is not something unique to technology. In fact the comparison to normal life is clearly visible: stay in public areas when with people you do not know, do not open mysterious boxes from people you do not know, and verify strange packages with people you do know. Of course you would not also put foreign items in your mouth for who knows where they have been! As they say abstinence is best, but if you are going to do it at least do it safely, and finally do not wear suggestive and provocative clothing unless you mean to be provocative and suggestive.

KNOWLEDGE

There are many temptations in life. If you want to be really good at what you do, you have to overcome the various temptations of life and focus on achieving your goals. If only you could be so lucky as to not work for an organization that thinks in years rather than experience. People often associate years performing a job to actual work experience and while this would seem to be the case, it is often not.

I know Network Engineers with 10 and 15 years of experience, who do not know the difference between port 23 and port 80. Of course, IP was never as important to networking as it is today. However, these people are those that never put forth the effort to do a little more than they need to. As such, they never learn more about their field than they need to in order to get by in their daily grind.

On the flip side, I know people with less than three years experience that could efficiently run a 200 node corporate network. Do not judge people because they have had years to learn something. Judge people on what they have learned in those years of experience. Remember that saying; don't judge a book by its cover.

Never be afraid to say, "I don't know". Pretending to have the answers to everything can be a bad career move. Of course, if you do not know, be sure to find out the answer. Not only will it help you broaden your knowledge, but it also shows commitment on your part to be the best at your job.

Remember to practice what you preach. Do not expect people to follow recommendations or guidelines you set if you do not follow them yourself. Lead by example and others will follow. If you want to succeed and become respected in your field, remember that there is always more to be learned. Seek it out and learn it. Become devoted to a life of learning.

There is a term representing those devoted to a life of learning, unfortunately the term has taken on a negative connotation and has been misused. The continual strive of learning is the truth behind real hackers. Hackers are the masters of their domain, whether NT or otherwise. Network Engineers, and other IT professionals, should also be hackers and strive to learn all they can about their fields. In the same respect, medical doctors are hackers as well. Doctors strive to learn all about a certain field of medicine, making them experts in their field. There is nothing special about the name hacker that binds it to the world of technology. Any person who is truly committed to a life of learning could be classified as a hacker.

Hackers do not use knowledge for evil. True hackers seek out defects or other holes in technology and notify people to fix them in order to benefit everyone by the forward progression of technology. However there is another type of hacker that has plagued the technological landscape. These pseudo-hackers use knowledge for evil. They seek out defects to crack holes in technology and exploit them for personal gain. There is nothing glamorous about these people. Unfortunately, the media misrepresents these people as hackers too.

There is no, "it doesn't work". As a technology professional, you find a way to make things work. It may not look pretty, but it gets the job done. You should take pride in your work and make its appearance as pleasant as its performance, but the bottom line is that the end user sees performance. To help you reach your goals it is important to remember that sometimes the best things in life are free.

The best network utility programs I have ever used have been free. Typically these programs were created by other Network Engineers to solve a particular problem. As such, these utilities gather the exact information that a Network Engineer needs. Open source code can be a lifeline for a Network Engineer who needs a special utility to get a job done. The wonderful thing about open source is many times you do not need to reinvent the wheel, but rather refine it for your purpose. Many times someone has already encountered the same problems as you. If you can find that person or resource it can make life much easier. Thank goodness for open source code.

As you proceed in life remember your knowledge is your own. The company does not own what you have learned. They have paid for your classes and all your certifications, but you have repaid them with your service. Of course, it is not a very smart resume builder to use a company just for training and then leave. And yet on the flip side, it is not smart to hang around with a company in a boring position just because you feel you owe them. You should do what will help you achieve your goals.

All too often, certifications are often misrepresented both by those that earn them and those who see them. Certifications should be used as a means of judging the abilities of someone against another. However, certifications should not be used as the means of identifying knowledge. What is important is not how many extra letters are after a person's name, but what the person knows and their ability to express that knowledge. Certifications are just letters next to a name.

When someone dictates a certain way to do things, do not be afraid to question. Your goal should be to better the company through the right information technology design. Ideas are often just thrown together and implemented without anyone ever really questioning if it was the right thing to do or not. Question the design. If there is solid reasoning behind the design then all is good. If not, you just saved the company from making a mistake and wasting valuable time.

The life of a technology professional is a life unrewarded. There is a misconception that Network Engineers do not really do anything. We rarely receive credit for keeping the network running, but we are the first to blame when there is a problem. To make matters worse, from a management perspective, there is no real measure to determine the productivity of a Network Engineer.

Our services are not like haircuts that can be counted and rewarded by the number completed. If a Network Engineer is really good, his or her services may go unnoticed for sometime. In fact, it can be argued the best Network Engineer is the one the company does not even know it has. If there is never a problem with the network then the Network Engineer must be doing something right. To learn we must question everything.

I almost forgot to mention the one thing you should never question. If you travel to Las Vegas for a business conference, remember to tip the airport shuttle driver. Yes, I know he only handled your bags for a few milliseconds, but he may just make an audible insanity plea that would make even Stella jealous.

As you have read, the life of a networking professional can be fun and rewarding. The continual advancements in IT and the networking field mean there will always be something new to learn. Organizations offer their own unique twist to the use of technology, and the Network Engineer ensures integration is successful.

As we approach the final chapter of this book we are merely beginning to explore the relationships and interactions between technology and people. We dream about the benefits to be gained in the future through the acceptance of new technologies. Yet, the future is an unmarked slab of stone and what will be written upon it remains unknown. Only time shall tell.

5

Only time shall tell

There once was a time when shear manpower ruled the economic world. Those glory days no longer have the dramatic impact they once did on the economic world. The present calls for the intelligent use of computer-power, an underlining theme of the so-called Information Technology ID-10T Effect. The ID-10T effect was directly born as a result of the increasing importance of computing power and information sharing. The world of computers has evolved, from the inventive make-it-yourself computers to the nanosecond super computers of yesteryear.

The world of today relies on large amounts of information accessible at very high speeds. An increase in the economic value of information has allowed for greater productivity. However, society must weigh the cost of protecting an individual's privacy to the cost of jobs and productivity.

In a world based on the theory of natural selection and the survival of the fittest, society must adapt or face extinction. Deciding what measures to take in protecting information privacy is a choice in the process of adaptation or extinction. Society must decide these measures for privacy and against those of productivity. As the decisions are made, we all must accept the consequences that follow.

If society decides to accept the need for information over individual privacy, the individual is left at the mercy of society. Society may ultimately have the decision of determining the amount of information security an individual receives as well. The world of information may be a place filled with new advancements benefiting society as a whole.

THE FUTURE

Nanosecond plus speeds will relay information all across the globe. Barriers in communication will no longer exist in a world were everyone speaks the same language: the computer language. All the functions of interacting with others will be conducted electronically. Financial transactions are based on credit cards, debit cards, and smart cards. Paper currency transactions have become to slow and inadequate. Paper memos and contracts no longer exist in the electronic business world. All communications will be done by teleconferencing. Information will be stored on secondary storage media.

An individual of the global population receives one credit card size smart card containing all of the information related to that person, from birth date and current weight to credit history and last purchase. Anyone and everyone has access to everyone else's records. The world lacks individual privacy, but the lack of meaningless privacy can surely be outweighed by the benefits of the computer evolution.

From the advancement in computer technology and information systems, medical science has improved tenfold. In a world free from all disease, an estimated 200 billion people with an average age of 104 years live peacefully. The method of warfare has changed with the new technology. Wars no longer cause massive destruction, for now they are fought in the realm of virtual reality. The evolution of technology has allowed for an evolution of mankind as well.

Society has obviously advanced greatly from its free access to information. From a distance the world appears to revolve around the good of the many, instead of the good of the few or the one. In this society, the privacy of the individual does not exist and society as a whole has adapted to the new environment.

People have learned to take responsibility for their actions. No longer does society ridicule people for their differences. The world has evolved to a higher plane of human interaction. This could be the future of the world, but if society chooses not to adapt to change we may become extinct.

In the refusal of society to accept the need for more information, society rejects computers as a whole. Society's rejection of computers has resulted in a collapse of technology. The world has begun to retreat back to its glory days of man and mule.

International business suffers from a lack of world communication. The absence of a communication network bars and even slows commerce to a practically non-existent state. Isolationism has taken hold.

The environment suffers from a demand for more natural resources. Demand for paper has increased tenfold, leading to an extinguishing of all the rain forest and the eventual depletion of the world's natural resources.

People are afraid of the outdoors. It has become infested with disease and violence. Nations begin to collapse. Inefficient communication restricts the sharing of ideas and research to cures for widespread diseases. Wars are continuously fought, more violent and bloodier than ever. This new world is our own creation for having feared the unknown. We now must accept the fate from resisting the change of technology. We chose a lifestyle more fulfilling to our needs in the present, not of our future. We may have chosen wisely or blindly, but now society must endure the legacy of its decision.

A time shall come when we as a society will decide our future relationships with computers and the expansion of technology. Embrace or reject. Hopefully, we will choose a careful course of action when we make such a decision. A slow progression into the information age will surely allow for the appropriate laws to be created, protected, and enforced for protecting the rights to individual privacy. This dilemma represents the heart of the Information Technology ID-10T Effect, or more commonly called being an information technology idiot. Will the world be the IT idiot, all to slow in reading its network for dummies book? Or will we be smart. If we look back on the history of humanity and its choices, I unfortunately see a society that will be foolish before it will learn how to properly use the power at its fingertips. You never can tell if the stove really is hot, unless you touch it...a few dozen times.

A compromise between respecting the rights of the individual and the need for information will ensure that society is successful in the ever-changing world of tomorrow. Society may someday determine the real economic value of information, but only after that economic value is established can society decide the amount of information necessary to ensure productivity while still protecting the privacy of the individual. In a world where change comes like the tides of a great sea, only time will tell what society has planned for its future. Only the adventurous and foolish would dare travel beyond the network.

Here in this place we are safe. We keep are heads down and continue onward. Here is our existence, our world, and our life. We create our own reality. We do not look beyond, but embrace what we have at our fingertips. Is our life true or false? On or off? Be it a one or a zero it matters not, for today we live on the network. May the network be your guide.

References

Duncan, WJ. (1988). *Great Ideas in Management: Lessons from the Founders and Foundations of Management Practice.* Jossey-Bass Publishers.

Toffler, Alvin. (1980). *The Third Wave.* Bantam Books.

978-0-595-36494-7
0-595-36494-2

www.ingramcontent.com/pod-product-compliance
Lightning Source LLC
Chambersburg PA
CBHW021043180526
45163CB00005B/2258